NATURE
OPTIONS

GW00733362

by Barbara Sharp

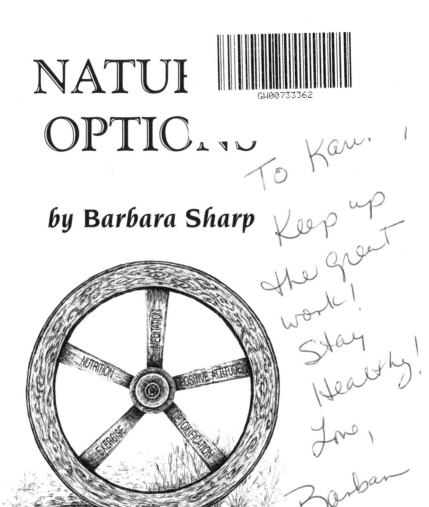

To Kau.
Keep up
the great
work!
Stay
Healthy!
Love,
Barbara
Sharp

A NATURAL
HEALTHCARE GUIDE

Illustrated by Christine D. Therrien

Natural Health Resource Centre, Inc.
Manchester, New Hampshire
1995

Natural Health Resource Centre, Inc.
P.O. Box 4195
Manchester, NH 03108-4195

To My Sons,

BRETT AND ADAM,

My Greatest Teachers

Thank you for your endurance
during my journey . . .

17 July 1995
Inis Maein,
Ireland

Dear Reader,

On this beautiful morning overlooking the Atlantic Ocean from the kitchen of my rental cottage on the Aran Islands, off the west coast of Ireland, I put pen to paper, the old-fashioned way. Many people have asked me to write a natural healthcare quick reference guidebook. If a question arises, no need to wait for the next class, radio or TV show, etc., just open the book! I am finally writing this book because of this interest as well as my firm belief in the re-education of people in the ancient wisdom that seems to have been lost to my generation, the "baby boomers."

What better place to write this book than here where many aspects of ancient life can still be seen and felt today. The people's closeness to nature and the sea are very evident here unlike other parts of the world. Since we are just raindrops trying to return to the sea, our source from which we emerged, I believe this to be the appropriate place for writing this kind of book.

We each are unique individuals walking our own pathways with different needs, desires, beliefs, lessons, etc. I hope each of you will find at least one option that may help to improve the quality of your life. If this occurs, then I will have indeed succeeded in achieving my purpose for writing this book.

Much health and happiness to each of you
                    as you travel on your path ......

            Barbara

ACKNOWLEDGMENTS

I'd like to thank the many people I've met along the way for their support, inspiration, courage, patience, and belief in me. There are so many but I'll just name a few because they all know who they are:

My teachers, most notably Dr. Deepak Chopra and Dr. Ann Wigmore, who have shared their incredible knowledge.

My friends, especially Luella Reed-Early, Lisa Gold, Lisa Kaufmann, Karl Lindblad, and Carol Robb Dodds, who have shared their wisdom with me. Also, Peter Randall, Dr. Grace Sullivan, Peggy Cote, Christine Therrien, and John Clayton for their excellent advice.

A very special thank you to my "sister" Michal Lachover and Paul Smith, whose incredible courage and strength while they were on this earth inspired me to follow my dreams.

My family, particularly Vern Sharp, because if it wasn't for him, this book may never have been written.

CONTENTS

INTRODUCTION

What is good health? In Western medical terms, it is the absence of detectable disease. In the ancient healing cultures, good health was defined as a sense of well-being in mind, body, and spirit.

Today people are taking back more responsibility for their wellness. One out of three Americans are using complementary medicine today. We're tired of the side effects of drugs as well as their high costs. We are all different, so therefore, how can one solution work for all? Weren't we always taught never to take someone else's medicine? Isn't that what we're doing when one medicine is prescribed for a "symptom" and not for the person?

As people become more responsible for their health and happiness, additional specific information is being requested. We have to make choices about our health. We need to learn more about our bodies and the natural options in caring for our bodies in order to make informed choices. The choices we make today, will create our tomorrows. The better our choices today, the better our tomorrows.

Taking a tip from history can be very helpful as well as time-saving. Why should we spend time 're-inventing the wheel'? Our ancestors had an abundance of knowledge when it came to taking care of themselves naturally. They used the tools they had at their disposal like foods, herbs, and each other.

Before our fast-paced world, healthcare information was passed down from one generation to another. We need to re-learn all those hard earned, ancient, natural methods of keeping ourselves happy and healthy before they are lost.

Modern technology has certainly saved many lives but it doesn't teach us preventive natural healthcare. Dealing with the causes of diseases rather than treating the symptoms is ultimately more beneficial. History has proven mind over matter and positive thinking are vital in maintaining good health. When we combine modern medical technology with the wisdom of our ancestors, then we can create better tomorrows.

"The Doctor of the future will give no medicine,
but will interest his patient
in the care of the human frame,
in diet and in the cause and prevention of disease."
Thomas Edison

The natural methods our ancestors used are low-cost and obviously, time-proven. There are few side effects, if any, because these methods and techniques enhance the body's natural healing mechanisms. The body can and does heal itself while we are in a state of relaxation, such as sleep. Also, our body can produce the best pharmaceuticals. We need to learn how to stimulate the production of these natural drugs. Since our body intuitively knows how to heal itself, we need to listen to the wisdom of our body.

"We carry within us,
the wonders we seek without us......"

Many believe that all diseases are symptoms of a weakened immune system. Therefore, it is important to learn the causes of a weak immune system and ways to improve it. Treating the body holistically (believing the mind, body, and spirit are of equal significance in maintaining good health), has been the standard philosophy for centuries by many healing cultures. Adopting that philosophy and mastering the art of self-healing, may help us be the best that we can be.

"Youth is a gift of nature; Aging is a work of art."

The information in this book is not new. We are just re-discovering it. It is a collection of information gathered from many books (see BIBLIOGRAPHY), studies, lectures, and discussions with experts from around the world. There is a vast amount of natural healthcare information available. This is merely a small portion mostly dealing with the basics. It is a menu for reaching and maintaining good health naturally in an easy reading and quick reference format. Hopefully, this book will inspire you and give you a place to begin your journey in natural healthcare. It could also be a stepping stone for those people who wish to further their own research.

Choose the options that feel right for you but start out slowly with 1 or 2 options. Your body reached this level over time, not overnight. Anything, whether it is healthy or unhealthy, when introduced too rapidly can shock the system. Introduce new options to your daily routine slowly and only when it feels comfortable.

After some time, if there is no change, try another option. Dr. Deepak Chopra states that 98% of our body cells are replaced every year, therefore, what makes us feel better now may have little or no effect next year. Also, since we are each different, what helps one person may not help another.

NATURAL HEALTHCARE BASICS

Today, there are people living high in the Himalayas in West Pakistan, called the Hunzas, who live to be 120—145 years old. They are happy, eat well, get plenty of exercise and rest, and appear to have great spiritual lives. There are other cultures around the world who also experience this longevity. Perhaps, we can re-learn from these supposed "backward people" how to live longer, happier lives in good health.

There are 5 basic areas in maintaining good health that seem to be common throughout history among the healing cultures of the world. They are nutrition, detoxification, exercise, positive attitude and meditation. They are the spokes in our Wheel of Good Health.

Wheel of Good Health

If one of the spokes is weak or broken, the wheel will wobble and break. Then, we will be unhappy and won't feel or do our best. It's important to keep each spoke strong so that the wheel will turn smoothly down our pathway of life.

These 5 basic areas are barely studied, if at all, in most western medical universities. The focus of the medical schools appears to be sick care or trauma care and not preventive healthcare. It's nice to know there are hospitals and doctors who can save our lives with modern medical technology after accidents, but fortunately accidents don't happen everyday.

NATURAL HEALTHCARE BASICS (con't.)

Where do we turn for everyday good healthcare methods if our doctors are not trained in these areas? We can take back our responsibility for our own preventive healthcare. We can re-educate ourselves in the wisdom of our ancestors by reading books such as this, attending natural healthcare classes and visiting natural healthcare professionals. Be careful in choosing your natural healthcare institutions and professionals. Be sure they practice what they teach. As always, if you're not totally comfortable with your selection, make another choice.

On the following pages, you will find options to help keep each spoke healthy and strong. When we pay attention to all spokes, we will be better able to handle the bumps on our path.

NUTRITION

"Your remedies shall be your food
and your food shall be your remedies."
Hippocrates

Today, it's common knowledge that it is vitally important to our health and happiness to choose carefully what we put in our mouths. Dr. T. Colin Campbell from Cornell University has studied the Chinese diet, which is centered on rice and vegetables, with little meat and dairy products. The apparent effect is phenomenally low rates of heart disease, obesity, and cancer. When Chinese women come to America and eat our diet, they develop breast cancer at a much higher rate than women in China. Dr. Dean Ornish has proven that diet can contribute to heart disease and the damage from the heart disease can be reversed with a special low-fat diet.

Is it in our best interests to be vegetarian or at least eat very small amounts of dairy and meat products? After reviewing the following chart, which shows how humans compare anatomically with other animals, are we physically equipped more like meat eaters or plant eaters?

"Nothing will benefit human health
and increase the chances for survival of life on Earth
as much as the evolution to a vegetarian diet."

Albert Einstein

Anatomy Comparison Chart

	Carnivores Meat Eaters (Dog, Wolf)	Herbivores Plant Eaters (Cow, Sheep)	Frugivores Fruit Eaters (Monkey, Ape)	Humans
Claws	Yes	No	No	No
Perspires Through	Tongue	Skin	Skin	Skin
Sharp, Pointed Front Teeth	Yes	No	No	No
Flat, Grinding Molars	No	Yes	Yes	Yes
Saliva*	Acid	Alkaline	Alkaline	Alkaline
Length of Intestines	3 x Body	10 x Body	12 x Body	12 x Body
Colon	Smooth	Convoluted	Convoluted	Convoluted

* Acid saliva makes it difficult to digest grains. Alkaline saliva helps to pre-digest grains.
** Short intestines allow rapidly decaying material to pass quickly to avoid toxicity.

Nutrition Options

The following is a list of nutritional options. All options under the "Best to Consume" list are recommended by many different sources. Some options under the "NOT Best to Consume" list have been proven not to be in our best health interests; others are questionable and therefore, may not be the best for us.

We each react differently to certain foods at various times. Choose your options according to how you feel and how your body reacts. Please introduce your choices to your diet slowly.

Best to Consume

Organic Foods (Chemical Free)

❒   Taste better

❒   More nutritious

❒   Our ancestors had higher quality foods because their soil was not chemically-laden nor depleted by overuse

❒   Cost decreasing due to supply/demand. It's better to pay now for healthy foods than pay later with illness.

*Refer to Coleman Natural Meat
        in RESOURCES

Herbs (Teas, Pills, Plants)

☐ Highest quality food known to man

☐ Historically, have proven extremely beneficial

☐ Commonplace for our ancestors to have herb gardens

☐ You can visit your local herbalist or read one of many herbal reference books to learn more specifics

*Refer to *Herb Bible*
     by Earl Mindell

*Refer to Nature's Sunshine Products
     in RESOURCES

Herbs (Teas, Pills, Plants) *(cont'd.)*

❐ Many physical challenges stem from the lack of certain nutrients in our diet. My uterine fibroid tumors disappeared after I included specific herbs in my diet. Here are a few of my favorites:

Blue Green Algae — It is the very basis of the entire food chain. It is a complex, whole food containing many vital nutrients that supports whatever systems are most out of balance. Can raise your energy level as well as lessen those "junk food" cravings.

*Refer to Cell Tech in RESOURCES

Echinacea — This is a wonderful herb that has been known to improve the immune system. I take it as soon as I feel I'm coming down with something.

Golden Seal — If an acute situation exists that requires an immediate boost to the immune system, historically, golden seal has proven beneficial. It should not be taken on a regular basis because it may be less effective when really needed.

Garlic — This is a natural antibiotic. It has been known to lower cholesterol. Many veterinarians recommend garlic to repel pests from pets and can work on humans, too! I take it daily.

Vitamins/Minerals

❑ Check brands for their purity

❑ Check brands for their ability to dissolve inside your digestive tract. Some brands will not dissolve and end up in the toilet bowl!

❑ Since dosages vary with individuals, consult your natural healthcare professional for the dosage which is best for you

❑ Multiple vitamin pills may not be in your best interest since everyone's body requires different vitamins/minerals. Take only the vitamins/minerals you require. The following is a list of some of the essential vitamins/minerals.

Vitamins/Minerals *(cont'd.)*

Vitamin C

> Dr. Linus Pauling recommends a minimum 6 grams/day of Vitamin C. The correct dosage for your body can be determined by starting with 500 mg and slowly increasing the amount of Vitamin C until you get diarrhea, then backing off to the previous amount before the diarrhea. This dosage may change over time since our bodies are constantly changing. Vitamin C can be used as a laxative. I take 1-3 grams/day of Vitamin C according to how I'm feeling.
>
> *Refer to How to Live Longer and Feel Better by Dr. Linus Pauling

Vitamin E

> Minimum of 400 IU's of this great antioxidant has been beneficial to most people. Vitamin E has brought relief to many menopausal women.

Vitamin B6

> Enhances the immune system; is a diuretic; reduces PMS symptoms; helpful in treatment of allergies, arthritis, asthma, and carpal tunnel syndrome.

Vitamins/Minerals *(cont'd.)*

Potassium

Deficiency can contribute to chronic fatigue, mental instability, muscle weakness, dry mouth, restlessness, depression, diabetes, blood pressure. I get potassium from bananas or the potassium mineral supplement called KM.

*Refer to *Everything You Always Wanted To Know About Potassium But Were Too Tired To Ask* by Dr. Betty Kamen

*Refer to Matol Botanical International in RESOURCES

Pycnogenol

A blend of special bioflavonoids that is an antioxidant 50 times more potent than Vitamin E and 20 times more potent than Vitamin C. Research shows that it can counteract inflammation, enhance circulation, lessen the effects of stress, and improve visual acuity.

Coenzyme Q10

Similar to Vitamin E but may be a more powerful antioxidant. It enhances the immune system; beneficial in aging, MS, obesity, candidiasis, diabetes, allergies, asthma, cancer, etc.

Chromium

Maintains stable blood sugar levels. Our soil and water are chromium deficient as well as the average American diet.

Raw Foods

❐ Best to be the majority of a diet

❐ Sprouts are especially nutritious because they are newly released from the nutrient-dense seed

*Refer to Steve Meyerowitz's books (The Sproutman)

*Refer to The Sprout House and Ann Wigmore Foundation in RESOURCES

❐ Full of fiber. Ancestors ate 50 grams/day of fiber. Perhaps due to processed foods, the average American eats only 11 grams/day. People can experience weight loss after increasing their fiber.

*Refer to *The New Facts about Fiber* by Dr. Betty Kamen

(See Fiber under DETOXIFICATION Options)

Raw Foods *(cont'd.)*

❏ High in enzymes and nutrients. Cooking foods above 105°F kills many of the enzymes and nutrients. When using our reserves of enzymes to digest cooked foods, we have fewer enzymes to repair cells to prevent disease and aging. Suggestion is to take enzymes when eating cooked foods.

*Refer to *Enzyme Nutrition*
     by Dr. Edward Howell

*Refer to Dr. Ann Wigmore's books

Homeopathy

❒ Gentle yet potent medicines made from natural substances that are safe and non-toxic

❒ Prescribed for over 200 years in very minute doses under the philosophy of "Like cures like" or the Law of Similar

❒ Reminds the body that it can heal itself

❒ 40-50% of European doctors use homeopathy

❒ Can help with most acute or chronic health problems

*Refer to *Everybody's Guide to Homeopathic Medicines* by Dr. Stephen Cummings & Dana Ullman

NOT Best to Consume

Non-Organic Foods

☐ Don't taste as well as organics (i.e. cardboard-tasting tomatoes)

☐ The chemicals are stored in our cells

☐ Some people have gotten sick after eating a chemically ladened banana because they touched the "meat" after handling the peel. Can also experience itchy, fuzzy mouth and nausea after eating non-organic bananas.

☐ Non-organic meat/chicken usually contains antibiotics, pesticides and adrenaline. The adrenaline is from the animals' fear created by its often inhumane, torturous slaughter.

   *Refer to *Diet for a New America*
      by John Robbins

☐ Clean non-organic food thoroughly with a special wash that can be bought at most health food stores. Also clean the work areas so as not to spread the toxins to other foods.

☐ Peel non-organic food because that is where the largest concentration of chemicals remain. Unfortunately, that's also the location of most of the nutrients.

Milk / Dairy Products

❐ We are the only species on earth that drinks another species milk

❐ Bovine Growth Hormone (BGH) is now in most milk/dairy products produced and sold in the U.S. BGH has been banned in Europe because of its negative effects on humans and cows.

❐ It's very possible that children are becoming resistant to antibiotics because they've continually had small doses of antibiotics in their milk/dairy/meat products

❐ Can cause mucous in the body because the body views it as a foreign matter and encapsulates it in mucous for protection. Many children who have chronic ear infections are better after all dairy products are eliminated from their diet and given acidophilus.

*Refer to *Don't Drink Your Milk*
by Dr. Frank Oski

Milk / Dairy Products *(cont'd.)*

❐ Alternative to butter or margarine is Better Butter— half butter/half vegetable oil

    *Refer to *Laurel's Kitchen*
       by Laurel Robertson

❐ Alternative to milk is soy or rice milk found at most health food stores

Iceberg Lettuce

❏ Very little nutritional value

❏ Full of bacteria

❏ Alternative is green leaf, romaine lettuce, etc.

Seafood

❏ One of the least regulated industries; therefore, many tainted fish due to few inspections

❏ Many people erroneously believe they have the flu after eating tainted fish

❏ It is recommended to only buy farm-raised fish with no fishy odor or odd taste

Aluminum

- ❏ Link between aluminum and Alzheimer's disease

- ❏ Avoid aluminum pots and pans

- ❏ Avoid deodorants containing aluminum. I use a Thai stone.

Coffee/Soda Drinks

- ❏ Because caffeine is a drug and alters the body's systems, it is best to avoid

- ❏ It is recommended to buy grain beverages like Roma or Inka

Fluoride

☐ This poison is used in rat and roach pesticides

☐ Proven ineffective as a decay deterrent by a study of 40,000 school children in 1986-1987 by the National Institute of Dental Research

☐ Can cause stained, discolored, or mottled teeth

☐ Linked to bone cancer in children and hip fractures in the elderly

☐ According to the U.S. Pharmacopoeia and the Physicians' Desk Reference, can cause hypersensitive reactions

☐ Poisons the immune, skeletal, and digestive systems

    *Refer to *Fluoride—The Aging Factor*
        by Dr. John Yiamouyiannis

☐ It has been recommended to use natural toothpastes that don't contain fluoride

    *Refer to Environmental Dental Association
        in RESOURCES

Dental Silver/Mercury Fillings

❐ Mercury is one of the oldest recognized poisons and is poisonous to all living cells

❐ Mercury vapors from mercury fillings can be released by chewing, drinking hot liquids, and brushing the teeth

❐ Can effect the nervous and immune systems causing depression, irritability, inability to concentrate, mental/emotional problems, headaches, fatigue, swollen glands, chronic infections, etc.

❐ Can be replaced with gold, porcelain, or resin fillings

❐ Be careful that an environmentally conscious dentist removes your fillings properly to lessen exposure to mercury vapors

*Refer to *The Complete Guide to Mercury Toxicity from Dental Fillings*
by Dr. Joyal W. Taylor

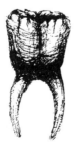

Healthy Eating Habits

How and when you eat is as important as what you eat. The following is a list of options to help you get the best out of your food.

☐ Eat slowly in a stress-free environment to ensure nutrient absorption. You may also find yourself eating less.

☐ Chew food at least 20-30 times to pre-digest food which aids your digestive tract

☐ Eat fruit alone because it digests in approximately 20 minutes while other foods can take 2-3 hours. If fruit is eaten with other foods, it will start to ferment while waiting for the other foods to digest which can disrupt your digestion with gas, etc.

☐ Eat before 7PM or 3 hours before bed to allow your body to finish digestion so that the body can focus on repairing and cleansing the cells during the night

Healthy Eating Habits *(con't.)*

❏   Antibiotics kill both good and bad bacteria. European doctors prescribe antibiotics with yogurt for its acidophilus. Take yogurt or acidophilus with antibiotics.

❏   Don't over eat because it can stress the digestive system. When food isn't digested properly, nutrients may not get absorbed.

❏   Drink fluids 15 minutes prior to or 15 minutes after eating so the fluid does not dilute the stomach digestive juices which can cause difficulty in digestion

❏   U.S. Surgeon General's Report stated 2/3 of all deaths in 1987 were related to diet!

DETOXIFICATION

Detoxification or toxin cleansing is the cleansing of the body and mind of impurities. Historically, our ancestors used fasts (which is one form of cleansing by not eating) to clean their minds prior to making important decisions.

Our world today is toxic. To be our healthiest, we need to rid ourselves of toxic air, food, water, and relationships. By living under stress, we manufacture our own toxins internally. Almost everyone can benefit from detoxification. As we spring clean our houses, we can spring clean our bodies and minds!

"A body free of toxins, mucous, poisons, dead cells, etc.
is stronger, healthier and has more energy."

Reasons for Cleansing:

Treat disease   Increase energy

Prevent disease   Increase creativity

Rejuvenation   Relax

Weight loss   Mental clarity

Clear skin   Slow aging

Aid our digestive organs, liver,
gall bladder, and kidneys

Signs and Symptoms of Toxicity:

Headaches   Allergies

Acne   Digestive problems

Sinus congestion   Insomnia

Frequent colds   Anxiety

Fatigue   Nervousness

Aches and pain   Depression

Nausea   Constipation

**"Every tissue in the body is fed by the bloodstream
which is supplied by the bowel.
When the bowel is dirty, the blood is dirty
and so are the organs and tissues.
It is the bowel that must be cared for first
before any effective healing can take place."**
Dr. Bernard Jensen

Let's begin with the colon which is one of the most important eliminative organs we have since 80-90% of all diseases begin here. What good is eating all the best foods, herbs, and vitamins if the nutrients are not getting through a clogged, constipated colon?

The ideal transit time of food entering the mouth to the toilet is 18-24 hours. The average American's transit time is 3 days! Visualize a plate of food sitting on your table for 3 days. Food putrefies inside of us, too!

Good bowel function entails having 2 or 3 good bowel movements per day. If we eat 3 meals a day and only have one bowel movement, where are the other 2 meals? What would our homes look like if we stopped taking the garbage out and kept the garbage inside? Our home would be like a clogged, constipated colon! Babies defecate after meals, why don't we anymore?

When we don't have regular bowel movements, mucoid fecal matter sticks to the colon walls. This can cause self-poisoning or auto-intoxification because the toxins are absorbed into the bloodstream. This mucoid fecal matter eventually becomes hardened and provides a great home for parasites and unhealthy bacteria. Some obese people are actually malnourished because the nutrients can't get through the colon walls and thus, their body is asking for more food/nutrients. After a colon cleanse, many obese people lose weight and their appetite declines because the nutrients are finally getting through.

My Story

My own personal story happened a few years ago when I was studying in Boston at the Dr. Ann Wigmore foundation. Dr. Ann had been surviving colon cancer for 35 years! She was in great shape and she was in her mid 80s! She believed raw foods, wheat grass juice, and colon cleansing kept her well with the cancer and told everyone who would listen! She passed on in 1994 from asphyxiation in a fire at the foundation but her legacy carries on.

While living at the foundation, we ate a very cleansing diet. Dr. Ann recommended daily enemas which I refused to do. After 3 days, my colon was overloaded with all the toxins being purged from my body. I could barely move; I felt very sick. Luckily, there was a colon hydrotherapist 5 blocks away so I very slowly and with much effort walked to her office for a colonic, which is a high enema. By the end of my one hour colonic, I was feeling great and I danced all the way back to the foundation! This experience was so dramatic and emphasized to me the importance of a clean colon.

Detoxification Options

Fasting

❏   Simple form of detoxification by not eating food

❏   Lower calorie consumption can pro-long life, lower blood pressure and cholesterol

❏   Fasting from 7PM to 7AM while the body is busy naturally cleansing has been recommended or fasting one full day/week

Water

☐   Water flushes the body of toxins

☐   Drink plenty of water (minimum 64 ounces/day)

☐   Warm distilled water is the most cleansing (for variety add a slice of lemon)

Fruit in the Morning

❑ By eating only fruit prior to noon, we can assist the body during its natural cleanse cycle

❑ Eating any other food during the natural cleansing cycle can inhibit cleansing while the body re-directs its energies to digesting

❑ Fruit is one of the most cleansing foods

*Refer to *Fit for Life*
   by Marilyn & Harvey Diamond

Rebounding (Bouncing on a Trampoline)

☐ The lymphatic system has valves that open and close during exercise (going against gravity) and carries away the wastes

☐ By bouncing on a trampoline or rebounding, every cell in our body is exercised

☐ Not necessary to bounce high or bounce aerobically to feel the benefits

☐ Some elderly and ill people start out sitting down or laying down on the trampoline while someone else does the bouncing. They can eventually work up to standing.

☐ Ideal to rebound for 20 minutes twice daily. You can do it while talking on the telephone or watching TV. Notice how good and energetic you feel afterwards!

*Refer to *The Cancer Answer*
by Albert E. Carter

*Refer to American Institute for Reboundology
in RESOURCES

Herbal Cleanses

❒ Certain herbs are great for cleansing the body and elimination of the wastes

❒ I take Cascara Sagrada daily to help my sluggish colon

❒ I have personally used 2 great herbal cleanses called Arise & Shine and the Purifying Program

*Refer to Arise & Shine and Eden Secrets
   in RESOURCES

Thymus Thump

❒ Tap the center of your chest just below your neck (breastbone) for approximately 20 seconds as often as you like

❒ This thumping stimulates the thymus gland and helps your immune system. Remember how an ape pounds his chest?

Fiber

❏ Lack of fiber has been linked to many diseases

*Refer to the chart entitled
   Diseases Caused by Lack of Dietary Fiber

❏ Recommended 30-50 grams/day

❏ If unable to get all the fiber in food, we can take fiber supplements. I take 1-3 tablespoons/day of ground psyllium powder in juice and water, ALWAYS followed with 8 ounces of water to ensure there will be no blockages. I started out with 1 tsp/day and increased slowly.

Fiber *(con't.)*

❐ Insoluble Fiber

A whole grain product that cannot be dissolved in water which acts like a broom in cleaning the digestive tract. The brushing effect and the decreased transit time lowers the concentration of bile acids. Examples are wheat bran, popcorn, dried bean seeds, rice bran, fruits, and vegetables eaten with the skin on.

❐ Soluble Fiber

Acts like a sponge in absorbing toxins and carcinogens along the digestive tract. Through absorption, this fiber flushes the system, lowers cholesterol, helps diabetics control blood sugar, control weight, and lower blood pressure. Mainly found in fruits, vegetables, barley, and oats.

*Refer to *The New Facts about Fiber*
    by Dr. Betty Kamen

*Refer to *Oat & Wheat Bran Health Plan*
    by Dr. Thomas & Dina Jewell

(See Raw Foods under NUTRITION—Best to Consume)

Diseases Caused by Lack of Dietary Fiber

Heart Disease

High Blood Pressure

Breast Cancer

Colon Cancer

High Cholesterol

Ovarian Cancer

Prostate Cancer

Hypoglycemia

Diabetes

Hemorrhoids

Constipation

Varicose Veins

Diverticulitis

Appendicitis

Weight Problems

Food Allergies

Colonics/Enemas

- ❐ Colon hydrotherapy (colonics) has been around since 1500 AD

- ❐ A colonic is an extended and more complete form of an enema. It involves gently infusing warm filtered water into the rectum through a tube. When the water flows out into the tube, it cleanses the colon of gas, mucous, infectious material, and feces.

- ❐ A colonic can be performed without discomfort to the client

- ❐ There is no smell or mess because it's a closed system of tubes connected to the drain

- ❐ Colonics and enemas are recommended as often as your colon needs cleansing. I have a colonic approximately once a month.

- ❐ Consult your natural healthcare professional or a colon hydrotherapist for more complete information

Squatting Position

- [ ] 80% of bowel cancers are found in the 2 areas of the colon that are not supported while sitting on a toilet

- [ ] This modern day position can cause kinked bowels as well as hinder complete elimination of fecal matter

- [ ] Squatting is the only position that promotes complete elimination of the bowel

- [ ] People have squatted throughout history and some are still doing so in different parts of the world today

- [ ] You can use a Welles Step or some books to prop up your legs to a squat position as seen below

*Refer to Welles Enterprises
   in RESOURCES

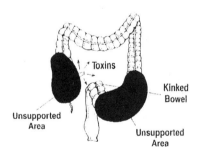

Massage

❒ Aids the lymph system in clearing our bodies of wastes, especially a specific massage called lymph drainage

❒ Brain chemicals that are produced in response to a massage improve our immune system and are still found in elevated levels in the body up to 3 weeks later

❒ Massage stimulates the production of interleukens which kill cancer cells

❒ PSAs (cancer indicators) can go down after a massage

Massage *(cont'd.)*

- ☐ Growth hormones increase after a massage—one of the reasons mother dogs lick their puppies!

- ☐ Massage is recommended after a heart attack because it can open the blood vessels to the heart

- ☐ Enhances the immune system, feels relaxing, and gives a sense of well-being

- ☐ You can give yourself a scalp massage or use one of those new wooden massage tools in between visits to your massage therapist

*Refer to *Perfect Health*
by Dr. Deepak Chopra

*Refer to *Ayurveda*
by Dr. Vasant Lad

Reflexology

- ❏ A 5,000 year old healing art originating in China

- ❏ A compression technique applied to specific points (reflexes) on hands, feet, and ears that unblock congestion in the corresponding organs and regions of the body

- ❏ It improves circulation, elimination, normalizes body rhythms, and reduces stress

- ❏ In many countries around the world, reflexology is practiced in hospitals and clinics. It is the most widely used form of complementary medicine in Denmark.

    *Refer to *Reflexology for Good Health*
    by Anna Kaye & Don C. Matchan

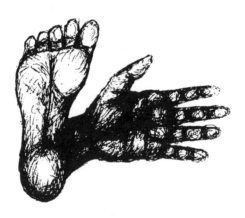

Breath

- ❐ Deep breathing rids our bodies of toxins both physically and mentally

- ❐ Diaphragmatic breathing (using the stomach) is very cleansing. Don't hold in your stomach—breathe deep!

- ❐ Babies breathe with their stomachs in a circular motion. We need to re-learn that method to enhance our immune system.

- ❐ It is recommended to practice deep breathing for approximately 1 hour/day

*Refer to *Perfect Health*
by Dr. Deepak Chopra

*Refer to *Ayurveda*
by Dr. Vasant Lad

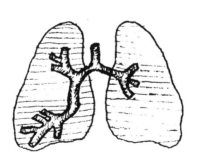

Skin Brushing

❏ Skin is our largest eliminative organ

❏ While we sleep at night, toxins rise to the surface of our skin

❏ The ancient science of Ayurveda recommends brushing the skin lightly with a soft brush or loofa sponge to remove the toxins in the morning

❏ It revitalizes the skin by removing the dead cells and toxins

*Refer to *Perfect Health*
    by Dr. Deepak Chopra

Meditation

- ☐ Cleansing the mind is of equal significance as cleansing the body

- ☐ Recommended 20-30 minutes twice daily

    *Refer to MEDITATION

Magnetic Therapy

- ☐ Magnets draw blood to the area where they are placed thus, enhancing the healing

- ☐ When you are feeling pain, take the magnets off your refrigerator and put them where you feel the pain!

    *Refer to *Magnetic Therapy*
    by Dr. H. L. Bansal

Chiropractic Care

❏ To keep the body functioning well, we need to keep it in line

❏ Visiting your local chiropractor twice a year or so helps to remove blockages. I visit my chiropractor with each change of the seasons.

❏ Chiropractors are trained in nutrition

*Refer to *The Truth About Chiropractic*
by Dr. Debra Levinson

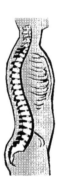

EXERCISE

In our society today, most everyone is aware of the importance of exercise. Yes, we do need to "use it or lose it," however, moderation is the key to good health. Any amount of exercise is better than no exercise because it helps to rid the body of toxins as well as strengthen all of the systems. One can still derive a sense of well-being from even a small amount of exercise which is where most people need to begin.

Start by finding something you enjoy doing and can have fun doing. If you don't like what you've chosen to do as exercise, you will have defeated the purpose, not to mention that the odds are against you maintaining this exercise for any length of time. If, after some time, you find you're not enjoying yourself with this particular form of exercise, change to another. Remember to start slowly at first and most important of all, have a good time!

The following are a few exercise options. Hopefully, you can find some form of exercise you like and can incorporate into your lifestyle with little effort. Where there is a will, there is a way!

Exercise Options

Walking

❒ One of the best exercises you can do

❒ It's fun and inexpensive however, it's very important to have proper walking shoes

❒ Slowly work up to walking 3 times/week for at least 30 minutes

Tai Chi/Chi Kung

❐ This ancient form of exercise is wonderful for your mind, body, and spirit

❐ It is a moving form of meditation

❐ It's relaxing and therapeutic, not to mention fun. Maybe, that's why it's been around for so many centuries!

❐ You can do this as often as you like

Yoga

- ❐   Yoga has been around for a long time

- ❐   Has similar benefits as Tai Chi

- ❐   You can do this as often as you like

    *Refer to *Ayurveda*
       by Dr. Vasant Lad

Climb Stairs

❏ If you have stairs, use them

❏ Or you can use a machine called the Stairmaster

❏ You can work up to 15-20 minutes 2 or 3 times/week

Dance

- [ ] Any kind of dance is great!

- [ ] Dance as often as you like!!

Horseback Riding

❐ This is very good exercise while you are communing with nature

❐ Do as often as you can

Aerobics

❑ Classes of varying intensities are offered in most communities

❑ 2 or 3 times a week for 30-60 minutes is recommended

Weight Training

❑ Either nautilus equipment or free weight training is fine

❑ Any age can participate. Elderly people were put on a simple weight program and had incredible results within 8 weeks!

❑ It's been proven to slow osteoporosis and increase bone density

❑ Lift a few for 20-30 minutes 3 times/week!!

Health Rider

❏   This latest machine is a great quick FULL body workout

❏   Three times/week for 20-30 minutes gets great results

*Refer to Exerhealth, Inc.
   in RESOURCES

POSITIVE ATTITUDE

The mind is the cornerstone of our health. We can consume healthy food, cleanse and exercise regularly but our minds can dirty our bodies again quickly. Our brain creates chemicals for each emotion. The power of thought is incredible.

"We are who we think we are."

Dr. Deepak Chopra says we can take poison and turn it into nectar with our minds. Also, negative thoughts can have a tremendous impact on our health. After a doctor tells someone they have 6 months to live (I could never understand how or why they would ever say that to someone), fifty percent of their immune system shuts down almost immediately! How's that for being powerful?

Another example of the power of the mind, is people who have Multiple Personality Disorders (MPD). One of the personalities may physically test positive for a disease, but when a different personality takes over the body, the person tests negative for the same disease! The same body may have the disease one day and not the next day, according to the state of mind.

Physiologically, our bodies cannot tell the difference between an experience that is actually occurring or a vivid imagination. The chemicals are being produced according to the emotion.

Our natural state is bliss which is being happy for no particular reason—just being alive. Being happy is usually for a reason like a pay raise or getting married.

"When I look for happiness, I lose it.
When I stop looking, and surrender to where I am,
I find it."

If you don't have a positive attitude perhaps you're ignoring the calling of your heart and desires. Are you living up to your full potential or are you afraid to try? Are you afraid of something?

"What if I try and I'm not very good?"

"What if I'm good and I don't try?"

"Our deepest fear is not that we are inadequate,
our deepest fear is that we are powerful
beyond measure.

It is our light, not our darkness,
that most frightens us.

We ask ourselves,
Who am I to be creative, talented, wise, loving?

Actually, who are you not to be?

You are a child of The Universe.

Our playing small does not serve the world.

There is nothing enlightened about shrinking
so that other people will not feel
insecure around you.

We are all meant to shine, as children do.

The light of The Universe is in everyone.

As we let our own light shine,
we give other people permission to do the same.

As we are liberated from our own fear,
our presence automatically liberates others."

Author Unknown

It is hard to have a positive attitude if we lack self-love. It is vital to our health and happiness to accept ourselves for who we are and love ourselves. If we love ourselves, we'll find love in everything around us and little will bother us. I'm sure you've had those days when you felt really great about yourself, everything went well all day and nothing annoyed you. Strive for more of those self-love days.

"What would it be like if you lived each day,
each breath, as a work of art in progress?
Imagine that you are a Masterpiece
unfolding every second of every day,
a work of art taking form with every breath."
Thomas Crum

Accept who you are, knowing that you are the best you can be today. Let go of the past and all its problems in order to live today in the present.

"Yesterday is history,
Tomorrow is a mystery.
Today is a gift,
that's why we call it the 'Present!'"

When you feel things are going wrong or you're facing what you perceive to be obstacles, remember that life isn't always what it appears. Our perception determines our feelings. Perhaps these obstacles are important lessons for us.

"... form serves us best when it works as an obstruction to baffle us and deflect our intended course. It may be that when we no longer know what to do, we have come to our real work and that when we no longer know which way to go, we have begun our real journey. The mind that is not baffled is not employed. The impeded stream is the one that sings."

Wendell Berry

"The wind blows strong and hard on tender young trees, not to harm them, but to teach their roots to hold firmly to the ground. I have grown into a tall, strong tree, with powerful roots. I bless the painful moments in my life, and I thank the wind. It's relentless force has molded me into the person I am today."

Dr. Barbara DeAngelis

ATTITUDE

"The longer I live, the more I realize the impact of attitude on life. Attitude, to me, is more important than the past, than education, than money, than circumstances, than failures, than successes, than what other people think or say or do. It is more important than appearance, giftedness or skill. It will make or break a company...a church...a home. The remarkable thing is we have a choice every day regarding the attitude we will embrace for that day. We cannot change our past...we cannot change the fact that people will act in a certain way. We cannot change the inevitable. The only thing we can do is play on the one string we have, and that is our attitude. I am convinced that life is 10% what happens to me and 90% how I react to it."

Chuck Swindoll

Positive Attitude Options

NLP (Neuro-Linguistic Programming) /Ericksonian Hypnosis

❐ This technology helps you to access your powerful inner resources to make your life even better by making positive changes

❐ Teaches better communication with each other

❐ Juggling an object between both hands is one technique used to end an anxiety attack. This activates the half of the brain that shuts down during an attack.

❐ Let your UNconscious be your guide!

*Refer to *Introducing Neuro-Linguistic Programming*
by Joseph O'Connor & John Seymour

*Refer to *My Voice Will Go With You*
by Sidney Rosen

*Refer to Creating Results, Inc.
in RESOURCES

*"If you keep doing what you've always done,
You'll always get what you've always got."*

Laughter

❑ Laughter is the best medicine

❑ Life is for fun. As Dr. Wayne Dyer says, "If you have a negative thought, say 'Next!'"

❑ Lighten-up and exercise your sense of humor! What ever makes you laugh—do it often!

*Refer to *Anatomy of an Illness*
    by Norman Cousins

Flower Essences

❏   "Nature's Prozac"

❏   Simple water-based extract that helps to resolve long-standing emotional problems

❏   Can relieve many physical complaints because it starts working on the thoughts or feelings that help fuel a physical problem such as trauma, depression, headaches, grieving, insomnia, psoriasis, paralysis, etc.

   *Refer to *Flower Essence Repertory*
       by Patricia Kaminski & Richard Katz

❏   Bach flower remedies are the most common products found at health food stores

   *Refer to Bach Flower Essences
       in RESOURCES

Light Therapy

❏ If you get depressed during the winter months when there is less sunlight, perhaps you suffer from Seasonal Affective Disorder (SAD)

*Refer to *Winter Blues*
    by Norman Rosenthal

❏ Some people have alleviated the depression by sitting under a special light (10,000 LUX box) for a few minutes every day

*Refer to The Sun Box Company
    in RESOURCES

Sound Therapy

❏ Our bodies create "happy chemicals" with different sounds

❏ Tranquilizing music has caused lymphocytes (immune system cells) that are inside the body and even outside the body in laboratory containers to create diazepam (natural valium tranquilizer)!

❏ Chanting music can create opiates in the body

❏ Listen often to whatever music makes you feel good and carry it with you to stressful situations

Aromatherapy

❏ Our bodies also create "happy chemicals" with different odors

❏ Specific odors can remind us of past events, both pleasant and uncomfortable memories

❏ Real estate agents can attest to how the probabilities in selling a house can increase when the owner is baking cookies or bread while the prospective buyer is viewing the house

❏ Discover your favorite smell and sniff often! Carry the odor with you on stressful occasions.

*Refer to Quantum Publications
   in RESOURCES

Express Feelings/Thoughts

☐ It is vital to our good health to share our feelings and thoughts ONLY with those people who are supportive

☐ To share with people who are not supportive is to set ourselves up for disappointment and frustration

☐ In 1975, the Spiegel Study found that women with metastatic breast cancer in support groups doubled their survival time

☐ In 1988, the Pennebaker Study found undergraduates improved their immune system function and reduced health clinic visits after writing about traumatic events

☐ Listening to or reading other people's experiences can help us to feel less alone

*Refer to *Chicken Soup for the Soul*
by Jack Canfield & Mark Victor Hansen

**"When we share our experiences with others,
we are amazed at the similarities!"**

Express Creativity

❏ Expressing your creativity can heighten your spirits tremendously

❏ The following exercises have been recommended:

> Write 3 longhand pages of your thoughts every morning. This is an active form of meditation and helps you to get in touch with your feelings.

> Make an artist date with yourself once a week to be adventurous, alone with yourself, engaging in an activity that is new, exciting and you've been wanting to do.

*Refer to *The Artist's Way*
by Julie Cameron

❏ If you stretch your demon to the absurd, it won't have any strength to bind you any longer. Alternate weeks with these different themes:

> Do only what you love, love everything you do
> Act as if everything is new
> Destroy judgment
> Ask dumb questions like "Why do cashiers stand?"

*Refer to *Creativity in Business*
by Michael Ray

"We come into this world alone,
and we leave alone.
In between, we must find whatever meaning
we can in our lives.
We must reach out to others and
celebrate with them these special times
in which we honor the past and
look with hope towards the future.
For it is only in these moments
that we transcend our human limitations,
break free from the bonds
of our solitary existence,
and taste the sweetness of life."

Author Unknown

Practice Non-Judgment

❐ Being judgmental doesn't allow us to have a positive attitude because it is limiting

❐ "Point of view" or "viewpoint" phrases are limiting

❐ When we do not judge ourselves and/or others, we free ourselves to accept everything as it is

❐ Non-judgment allows us time and energy to spend on positive endeavors

*Refer to *The Seven Spiritual Laws of Success* by Dr. Deepak Chopra

Sleep/Relaxation/Rest

❑   This is very basic but very important since the body heals itself both mentally and physically while in a relaxed state

❑   To deprive oneself of good rest is to invite physical and mental disorders

❑   Get as much sleep as you need to feel well rested

Symptoms of Inner Peace

Watch for signs of Peace. The hearts of a great many have already been exposed to it and it seems likely that we could find our society experiencing it in epidemic proportions. Some signs and symptoms of Inner Peace:

Tendency to think and act spontaneously rather than from fear.

An unmistakable ability to enjoy each moment.

Loss of interest in judging other people.

Loss of interest in judging self.

Loss of interest in interpreting the actions of others.

Loss of interest in conflict.

Loss of ability to worry (a very serious symptom).

Frequent, overwhelming episodes of appreciation.

Contented feelings of connectedness with others and with nature.

Frequent attacks of smiling through the eyes and from the heart.

Increasing tendency to let things happen rather than make them happen.

Increased susceptibility to Love extended by others as well as the uncontrollable urge to extend it.

If you have all or even most of the above symptoms, please be advised that your condition may be too far advanced to turn back. If you are exposed to anyone exhibiting several of these symptoms, remain exposed at your own risk. The condition of Inner Peace is likely well into its infectious stage. Be forewarned.

Author Unknown

MEDITATION

Prayer is asking questions of the universe, God, your higher self, or whomever. Meditation is listening for the answers. It helps us to get in touch with the power within us. There have been studies proving the physiological benefits from meditation such as:

Lowers blood pressure

Reduces stress and anxiety

Basal metabolic rate drops

Numerous psychosomatic disorders
are relieved and disappear

Improves attention span

Increases creativity

Enhances learning ability

Improves memory retrieval

"Taking time for silence means standing back far enough so you can determine if the pictures in your life are crooked or straight. Silence will help you see clearly, sometimes for the first time, exactly what is out of balance in your life."

Dr. Barbara DeAngelis

Meditation is a form of silence. You're not really forcing your mind to be quiet, you're accessing the quiet that is always there. When we're unhappy, feeling anxious or guilty, we're victims of memories and our internal dialogue is active. As Dr. Deepak Chopra explains in his book "Perfect Health," silence is the birthplace of happiness, where we get our bursts of inspiration, our tender feelings of compassion and our sense of love. The restful alertness experienced during meditation can be felt during the rest of the day.

It is recommended to meditate for 20-30 minutes twice daily. The best times seem to be early morning and late afternoon. Be careful if you meditate in the evening. You may feel so much energy that you may have difficulty falling asleep.

*"If you aren't used to going within,
you will need to be patient with yourself
while you learn how to navigate inner space."*

Dr. Barbara DeAngelis

Meditation Options

Primordial Sound

❏ Primordial sounds are nature's basic vibrations

❏ An appropriate sound or mantra for you is calculated based on the location, time, and date of your birth

❏ When used correctly, these mantras create a quiet, soothing effect in the mind

❏ While sitting comfortably and quietly in a peaceful place, continually repeat your mantra. Your thoughts may interrupt or you may feel restless, but that is stress being released like bubbles rising to the surface and bursting. Effortlessly, return to your mantra for the duration of your meditation time knowing that you will be more relaxed.

*Refer to Quantum Publications
    in RESOURCES

Transcendental

❏ Reciting a mantra or meaningless word to help your conscious mind focus on something other than internal dialogue

Visual

❏ Some people feel calm while looking at the ocean, forest, a favorite picture, etc. Whatever calms you, look at it.

Sound

❏ If listening to the sounds of nature, chanting music, etc. soothes your soul, you can incorporate those sounds during your meditation time

Active

❏ Tai Chi, Yoga, writing, walking, etc. can be a form of active meditation if you feel peace

CONCLUSION

*"If you don't take the time to be well,
you WILL take the time to be ill!"*

At the end of this book, there is a bibliography and a list of resources for the books, products and people mentioned throughout these pages. This book is intended for informational purposes only. Please consult a natural healthcare professional before changing any aspect of your life.

As you improve your diet and health, you might experience some puzzling symptoms. I've included an excerpt from an article explaining what you may experience. After the excerpt are two of my favorite inspirational messages. I hope you enjoy them and learn from them as I have. Until we slip out of the harbor of our body and sail back into the realm of spirit, I wish you health and happiness............

DIET IMPROVEMENT SYMPTOMS

Perhaps the greatest misunderstanding in the field of nutrition is the failure to understand and interpret the symptoms and changes which can occur at the beginning of a better nutritional program. A remarkable thing happens when a person improves the quality of the food they consume.

When the food you ingest is of higher quality then the tissues from which the body is made, the body discards the lower quality tissues, to make room for the higher quality materials to make healthier tissue.

During this process of regeneration, lasting about 10 days to several weeks, the emphasis is on breaking down and eliminating lower quality tissue. The vibrant energy found in the external parts of the body, the muscles and skin, moves to vital internal organs and starts reconstruction. This movement of energy produces a feeling of less energy in the muscles, which the mind interprets as weakness. At this time, more rest and sleep is often needed. And it's imperative to avoid stimulants of any kind which will abort and defeat the regenerative process. Remember the body isn't getting weaker. It's simply using it's energies in more important internal work, rather than external work involving muscle movements. With patience and diligence, a person will soon feel more energy than before.

By ingesting higher quality foods, the body begins a process called "retracing." The initial focus is on eliminating the waste and toxins deposited in the tissues. However, the process creates symptoms that are often mis-interpreted.

For example, a person stops consuming coffee or chocolate and experiences headaches and a general letdown. The body begins discarding toxins (caffeine or theobromine) by removing them from the tissues and transporting them through the bloodstream. However, before toxins are passed, through elimination,—they register in our consciousness as pain—in other words a headache. These same toxins also stimulate the heart to beat more rapidly, thus producing the feeling of exhilaration. The letdown is due to the slower action of the heart, which produces a depressed state of mind.

The symptoms experienced during "retracing" are part of the healing process! They are not deficiencies. Do not treat them with stimulants or drugs. These symptoms are constructive even though unpleasant at the moment. Don't try to cure the cure.

The symptoms will vary according to the materials being discarded, the condition of the organs involved in the elimination, and the amount of available energy. They can include:

- ❏ Headaches
- ❏ Fever/Chills
- ❏ Colds
- ❏ Skin Eruptions
- ❏ Constipation
- ❏ Diarrhea
- ❏ Fatigue/sluggishness
- ❏ Nervousness
- ❏ Irritability
- ❏ Depression
- ❏ Frequent Urination

The symptoms will be milder and pass more quickly, if one gets more rest and sleep. Understand that the body is becoming healthier by eliminating waste and toxins. Had they remained trapped in the tissues, eventually they would have brought about illness and disease, thus causing greater pain and suffering.

Finally, don't expect to improve your diet and feel better and better every day, until you reach perfection. The body is cyclical in nature. Health returns in a series of gradually diminishing cycles. For example, you may begin eating better and start feeling better. After some time, you experience a symptom such as nausea, or diarrhea. After a day, you feel even better than before and all goes well for a while. Then, you suddenly develop a cold, the chills, and lose your appetite. Without the use of drugs, you recover from these symptoms and suddenly feel great. This well-being continues for a time until you break-out in a rash. The rash flares-up, but finally disappears. And suddenly you feel better than you've felt in years.

As the body becomes pure, each reaction becomes milder and shorter in duration; followed by longer and longer periods of feeling better than ever before, until you reach a level plateau of vibrant health.

by Dr. Stanley S. Bass, N.D., D.C., Ph.C.

LESSONS FROM THE GEESE

As each bird flaps its wings, it creates an uplift for the bird behind it. By flying in formation, the birds can fly 71% farther than if they fly alone.

The Lesson: People who share a common sense of purpose can get where they want to go quicker and easier when they are propelled by the thrust of others who share the same goals.

Whenever a goose falls out of formation, it suddenly feels the difficulty of trying to fly alone and quickly gets back into formation.

The Lesson: We should stay in formation with those who are headed in the direction we want to go. We can accomplish much more together than by ourselves.

When the lead goose gets tired, it rotates back into the formation and another goose flies at the point position.

The Lesson: It is only fair that we take turns doing the hard tasks and share leadership responsibilities. Leaders must have followers and followers need leaders.

The geese in formation honk from the rear to encourage those up front to keep on course and maintain their speed.

The Lesson: We need to make sure our honking from behind is helpful and encouraging. Let's remember to say please and thank you to those we serve.

When a goose gets sick or shot down, two geese drop out of formation and follow him down to help and protect him. They remain with him until he dies or is able to fly again.

The Lesson: It is easy to like those who are like us. However, our true character is revealed in our response to those around us who are hurting or suffering misfortune.

Author Unknown

IF I HAD MY LIFE TO LIVE OVER

"I'd dare to make more mistakes next time. I'd relax. I would limber up. I would be sillier than I have been this trip. I would take fewer things seriously. I would take more chances. I would take more trips. I would climb more mountains and swim more rivers. I would eat more ice cream and less beans. I would perhaps have more actual troubles, but fewer imaginary ones. You see, I'm one of those people who live sensibly and sanely hour after hour, day after day. Oh, I've had my moments. If I had it to do over again, I'd have more of them. In fact, I'd try to have nothing else. Just moments, one after another, instead of living so many years ahead of each day. I've been one of those persons who never goes anywhere without a thermometer, a hot water bottle, a raincoat, and a parachute. If I could do it again, I would travel lighter than I have. If I had my life to live over, I would start barefoot earlier in the spring and stay that way later in the fall. I would go to more dances. I would ride more merry-go-rounds.

I would pick more daisies."

Nadine Stair and Elizabeth Lucas

SUMMARY OF OPTIONS

NUTRITION

Best to Consume
- Organic Foods
- Herbs
  - Blue Green Algae
  - Echinacea
  - Golden Seal
  - Garlic
- Vitamins/Minerals
  - Vitamin C
  - Vitamin E
  - Vitamin B6
  - Potassium
  - Pycnogenol
  - Coenzyme Q10
  - Chromium
- Raw Foods
- Homeopathy

Not Best to Consume
- Non-Organic Foods
- Milk/Dairy Products
- Iceberg Lettuce
- Seafood
- Aluminum
- Coffee/Soda Drinks
- Fluoride
- Dental Silver/Mercury Fillings

Healthy Eating Habits

DETOXIFICATION

Fasting
Water
Fruit in the Morning
Rebounding (Trampoline)
Herbal Cleanses
Thymus Thump
Fiber
> Insoluble Fiber
> Soluble Fiber
Colonics/Enemas
Squatting Position
Massage
Reflexology
Breath
Skin Brushing
Meditation
Magnetic Therapy
Chiropractic Care

EXERCISE

Walking
Tai Chi/Chi Kung
Yoga
Climb Stairs
Dance
Horseback Riding
Aerobics
Weight Training
Health Rider

POSITIVE ATTITUDE

NLP (Neuro-Linguistic Programming)/
    Ericksonian Hypnosis
Laughter
Flower Essences
Light Therapy
Sound Therapy
Aromatherapy
Express Feelings/Thoughts
Express Creativity
Practice Non-Judgment
Sleep/Relaxation/Rest

MEDITATION

Primordial Sound
Transcendental
Visual
Sound
Active

BIBLIOGRAPHY

Heart of the Mind by Connirae & Steve Andreas
    Real People Press

Prescription for Nutritional Healing by James F. Balch, M.D.
    Avery Publishing Group

Magnetic Therapy by Dr. H. L. Bansal
    DeVorss & Co.

The Foot Book by Devaki Berkson
    Harper Perennial

Who Said So? by Rachelle Breslow
    Celestial Arts

Alternative Medicine by The Burton Goldberg Group
    Future Medicine Publishing, Inc.

The Artist's Way: A Spiritual Path to Higher Creativity
    by Julie Cameron
    G.P. Putnam's Sons

Chicken Soup for the Soul (and A 2nd Helping of)
    by Jack Canfield and Mark Victor Hansen
    Health Communications, Inc.

The Cancer Answer by Albert E. Carter
    A.L.M. Publishers

Body Reflexology by Mildred Carter
    Parker Publishing Company

Ageless Body, Timeless Mind by Dr. Deepak Chopra
    Bantam Books

Creating Health by Dr. Deepak Chopra
    Houghton Mifflin Co.

Bibliography (continued)

Journey Into Healing by Dr. Deepak Chopra
  Harmony Books

Perfect Health by Dr. Deepak Chopra
  Harmony Books

Quantum Healing by Dr. Deepak Chopra
  Bantam Books

The Seven Spiritual Laws of Success by Dr. Deepak Chopra
  Amber-Allen Publishing

Anatomy of an Illness by Dr. Norman Cousins
  Bantam Books

Everybody's Guide to Homeopathic Medicines
  by Dr. Stephen Cummings & Dana Ullman
  Jeremy P. Tarcher, Inc.

Real Moments by Dr. Barbara DeAngelis
  Delacorte Press

Fit For Life by Harvey & Marilyn Diamond
  Warner Books

Real Magic by Dr. Wayne Dyer
  Harper Collins Publishers

The Complete Guide to Mercury Toxicity from Dental Fillings
  Environmental Dental Assoc.

Third Opinion by John M Fink
  Avery Publishing Group, Inc.

The Goldbecks' Guide to Good Food by Nikki & David Goldbeck
  New American Library

Bibliography (continued)

A Holistic Protocol for the Immune System
by Scott J. Gregory, O.M.D.
Tree of Life Publications

World Without Cancer by G. Edward Griffin
American Media

You Can Heal Your Life by Louise Hay
Hay House

What Your Doctor Won't Tell You by Jane Heimlich
Harper Perennial

The New Holistic Herbal by David Hoffmann
Element Books Ltd.

Enzyme Nutrition by Dr. Edward Howell
Avery Publishing Group

Tissue Cleansing Through Bowel Management
by Dr. Bernard Jensen
Jensen Enterprises

Oat & Wheat Bran Health Plan by Dr. Thomas & Dina Jewell
Bantam Books

*Everything You Always Wanted To Know About Potassium
But Were Too Tired To Ask* by Betty Kamen, Ph.D.
Nutrition Encounter, Inc.

The New Facts about Fiber by Betty Kamen, Ph.D.
Nutrition Encounter, Inc.

Flower Essence Repertory by Patricia Kaminski & Richard Katz
Flower Essence Society

Reflexology for Good Health by Anna Kaye & Don C. Matchan
Wilshire Book Company

Bibliography (continued)

Ayurveda—The Science of Self-Healing by Dr. Vasant Lad
  Lotus Press

The Truth About Chiropractic by Dr. Debra Levinson
  Max Publications

Growing Vegetables Indoors by Steve Meyerowitz
  The Sprout House

Recipes from The Sproutman by Steve Meyerowitz
  The Sprout House

Herb Bible by Earl Mindell
  Simon & Schuster

World Medicine by Tom Monte
  G.P. Putnam's Sons

Encyclopedia of Natural Medicine
  by Michael Murray, N.D. and Joseph Pizzorno, N.D.
  Prima Publishing

Introducing Neuro-Linguistic Programming
  by Joseph O'Connor & John Seymour
  Mandala

Reversing Heart Disease by Dr. Dean Ornish
  Harper Collins Publishers

Don't Drink Your Milk by Dr. Frank A. Oski
  TEACH Services

Homeopathic Medicine at Home
  by Dr. Maesimund B. Panos & Jane Heimlich
  Jeremy P. Tarcher, Inc.

How to Live Life Longer and Better by Dr. Linus Pauling
  Avon Books

Bibliography (continued)

Creativity in Business by Michael Ray
    Doubleday

Diet for a New America by John Robbins
    Stillpoint

Laurel's Kitchen by Laurel Robertson
    Blue Mountain Center of Meditation

Illustrated Encyclopedia of Herbs by Rodale
    Rodale Press

My Voice Will Go With You
    —The Teaching Tales of Milton Erickson
    by Sidney Rosen
    W. W. Norton Company

Winter Blues by Norman Rosenthal
    Guilford Press

Bach Flower Therapy by Mechthild Scheffer
    Healing Arts Press

Love, Medicine & Miracles by Bernie S. Siegel, M. D.
    Harper Perennial

Diet for A Poisoned Planet by David Steinman
    Harmony Books

Hunza Health Secrets by Renee Taylor
    Keats Publishing, Inc.

Health Handbook by Louise Tenney
    Woodland Books

Nutritional Guide by Louise Tenney
    Woodland Books

Bibliography (continued)

Today's Healthy Eating by Louise Tenney
Woodland Books

Today's Herbal Health by Louise Tenney
Woodland Books

Many Lives, Many Masters by Dr. Brian L. Weiss
Simon & Schuster

Be Your Own Doctor by Dr. Ann Wigmore
Avery Publishers

Be Your Own Healer by Dr. Ann Wigmore
Ann Wigmore Press

The Sprouting Book by Dr. Ann Wigmore
Avery Publishers

Why Suffer? by Dr. Ann Wigmore
Avery Publishers

Fluoride—The Aging Factor by Dr. John Yiamouyiannis
Environmental Dental Assoc.

The China Study by T. Colin Campbell, Ph.D.
Cornell University,
Syracuse, NY

Pennebaker JW, Kiecolt-Glaser JK and Glaser R: Disclosures of Traumas and Immune Function: Health Implications for Psychotherapy. *Journal of Consulting and Clinical Psychology,* 1988; 56:239-245

Speigel D, Bloom JR, et al: Effect of Psychosocial Treatment on Survival of Patients with Metastatic Breast Cancer. *Lancet,* 1989; 2:888-891

Bibliography (continued)

New Age Journal Magazine
New Age Publishing, Inc.

Natural Health Magazine
Boston Commons Press Limited Partnership

RESOURCES

American Institute of Reboundology
    60 E. 100 South Ste 201
    Provo, UT 84606
    801/377-0570
        (Rebound Information and Trampoline)

Arise & Shine Colon Cleanse
    3225 N. Los Altos Ave.
    Tuscon, AZ 85705
    800/688-2444

Bach Flower Essences
    800/433-7523

Cell Tech (Blue Green Algae)
    1300 Main St.
    Klamath Falls, OR 97601-5900
    800/800-1300

Dr. Deepak Chopra
    Center for Mind Body Medicine
    973B Lomas Sante Fe Drive
    Solana Beach, CA 92075
    619/794-2425

Coleman Natural Meat
    5140 Race Court
    Denver, CO 80216
    303/297-9393

Creating Results, Inc.
    427-3 Amherst St.
    Nashua, NH 03063
    603/880-7765 or
    508/692-0405
        (NLP)

Eden Secrets
270 N. Cannon Dr. Ste 1414
Beverly Hills, CA 90210
800/952-7873
(Purifying Program Colon Cleanse)

Environmental Dental Association
9974 Scripps Ranch Blvd.
Ste 36; San Diego, CA 92131
619/586-1208

Exerhealth, Inc.
3616 West 1820 South
Salt Lake City, UT 84101
800/235-4455
(Health Rider)

G. Edward Griffin
American Media
PO Box 4646
West Lake Village, CA 91359
800/282-2873

Matol Botanical International
1111 46th Avenue
Lachine, Quebec, Canada H8T 3C5
800/363-1890
(KM)

Nature's Sunshine Products
P. O. Box 1000
Spanish Fork, Utah 84660
800/223-8225 or
800/453-1422

Quantum Publications
    PO Box 1088
    Sudbury, MA 01776
    800/757-8897
        (Primordial Sound Meditation/Aromatherapy)

The Sprout House
    40 Railroad St.
    Great Barrington, MA 01230
    413/528-5200

The Sun Box Company
    800/548-3968
        (Seasonal Affective Disorder (SAD)
        10,000 LUX Light Box)

Welles Enterprises
    6565 Balboa Ave. Ste. A
    San Diego, CA 92111
    619/473-8011
        (Welles Step)

Ann Wigmore Foundation
    PO Box 140
    Torreon, NM 87061
    505/384-1017 or 617/267-9424

ABOUT THE AUTHOR

Barbara Sharp, President of the Natural Health Resource Centre in Manchester, New Hampshire, is a lecturer, consultant, and researcher in natural healthcare for more than 25 years with an office in Toronto, Ontario, Canada. Barbara discusses Dr. Deepak Chopra's books, philosophy, and concepts in the Barnes & Noble bookstores in New Hampshire and is an Ambassador for the Greater Manchester Chamber of Commerce. She has studied with Dr. Deepak Chopra, Dr. Ann Wigmore, Jim Rohn, and many others. She is certified in NLP/Ericksonian Hypnosis, Reiki, and Applied Kinesiology.

Most of the information she shares with people has been virtually forgotten in the United States, thus she travels internationally to conduct research as well as lecture. Her firm belief in combining modern medical technology with the wisdom of our ancestors to create health and happiness is the focus of her educational seminars on TV, radio, and group presentations.

To arrange for a seminar contact:

Barbara Sharp
Natural Health Resource Centre, Inc.
P.O. Box 4195
Manchester, NH 03108-4195
(603) 471-0094